I0045046

Jacob Mendez DaCosta

Inhalations in the Treatment of Diseases of the Respiratory Passages

Particularly as effected by the use of atomized fluids

Jacob Mendez DaCosta

Inhalations in the Treatment of Diseases of the Respiratory Passages
Particularly as effected by the use of atomized fluids

ISBN/EAN: 9783337182250

Printed in Europe, USA, Canada, Australia, Japan

Cover: Foto ©berggeist007 / pixelio.de

More available books at **www.hansebooks.com**

IN THE

TREATMENT OF DISEASES

OF THE

RESPIRATORY PASSAGES,

PARTICULARLY AS EFFECTED

By the Use of Atomized Fluids.

BY

J. M. DA COSTA, M.D.,

Physician to the Pennsylvania Hospital; Fellow of the College of Physicians:
President of the Pathological Society of Philadelphia, etc. etc.

PHILADELPHIA:

J. B. LIPPINCOTT & CO.

1867.

Entered, according to Act of Congress, in the year 1867, by

J. B. LIPPINCOTT & CO.,

In the Clerk's Office of the District Court of the United States for the
Eastern District of Pennsylvania.

PREFACE.

THIS essay appeared originally in the September and October numbers of the *New York Medical Journal* of the past year. It was intended to set forth certain conclusions arrived at regarding inhalations; to be in fact an expression of personal experience on a subject which, though it had commenced again to attract some attention, was doing so rather in connection with the novelty of the apparatus used than from any clear ideas of its therapeutic importance. I began this inquiry in the spirit simply of examining the value of the plan as a clinical problem; and satisfying myself of its use in some disorders and its futility in others, published the results in the form alluded to, adding sufficiently, as I thought, of the *modus operandi* to

allow every physician to test the matter for
himself. The essay, of which some copies were
struck for private circulation, seems to have
attracted much attention, and the copies that I
had becoming soon exhausted, I have been re-
quested by my publishers to reissue it to supply
the demand that is made on them. It stands in
substance as it was published; here and there
some additions have been made.

J. M. Da Costa.

1005 Spruce Street, Philadelphia.
        February, 1867.

CONTENTS.

———— ——

CHAPTER I.

(v)

INHALATIONS

IN THE

TREATMENT OF DISEASES

OF THE

RESPIRATORY PASSAGES.

CHAPTER I.

THE HISTORY OF INHALATIONS AND THE APPARATUS EMPLOYED.

THE chief materials for this essay were gathered during my term of service at the Pennsylvania Hospital, though prior to it I had frequently used inhalations in private practice in laryngeal and pulmonary affections. Some of the results obtained—the negative as well as the positive—were communicated to the class attending the clinical lectures, and, indeed, many of the cases were seen by them. But as these results were mainly gained by a plan of comparatively recent origin, that of atomization, and one the value of which can only be settled by the conjoined labor of many, I have thought, whether rightly or wrongly, that they might be of interest to a larger circle than that which witnessed them, and contribute something toward the knowledge of the profession on the subject. Let me further premise, that a number of the cases which will serve as the basis of much which will be said in this essay, especially those under my charge at the hospital, received no other treatment, and thus I sought to avoid a stumbling-block in therapeutic reasoning which often

2   (9)

interferes so greatly with our efforts to arrive at definite conclusions.

The attempt to use inhalations in the treatment of disease is, as is well known, not a novelty. Aretæus employed them, and with Galen they were a favorite mode of treatment. Indeed, with all Roman physicians they became so; and both by the physicians of antiquity and later by the Arabs the inhalation of sulphurous vapors in bronchial affections, and of arsenical vapors in asthma, was constantly advised—the latter a prescription which, in our own times, has been revived by Trousseau. Hot, dry air was strongly recommended for the cure of consumption by Piso, in 1580; and Bennet, a London physician, living about the middle of the seventeenth century, directed his patients to breathe the atmosphere of a chamber filled with fumes of medicinal substances, especially of the gum-resins. Several authors of the sixteenth and seventeenth centuries lauded camphor, amber, myrrh, sulphur, assafetida, and the balsams for the purposes of inhalation; and Benedict* employed them in consumptive diseases. Boerhaave and Van Swieten used inhalations in the early part of the eighteenth century; and Boerhaave† gives several quite explicit formulas. Mead‡ advises that the balsamic ingredients for inhala-

---

* Theatrum Tabidorum.   † Materia Medica.
‡ Monita et Precepta.

tion should be thrown on red-hot coals, and the fumes received through a proper tube directed to the windpipe, and praises highly the smoke thus conveyed into the lungs, when produced by balsam of Tolu.

But it does not appear that any of these endeavors to employ inhalations were particularly successful, and they had fallen into disrepute, when the detection of oxygen and other elementary gases led to the mode of treatment being revived by Beddoes. Yet the exaggerated statements with reference to its action, the uncertain effects, and the attempt to make inhalations serve the purpose of a panacea, produced again, very naturally, an utter want of confidence in them, which was only disturbed by the discovery in this century of iodine and chlorine. These agents were eagerly seized hold of by those physicians who had not lost all faith in inhalations,—and prominent among them we find Scudamore and Piorry,—partly with the view of acting on affections of the respiratory organs; partly because a better knowledge of physiology was teaching that we may make use of these organs to modify the blood, and thus alter the condition of the whole system. Though here, too, we find that a practical application of this view had already for some time been made; for Fracastori* records that the fumes of cinnabar were much employed by inhalation in the treat-

---

* See Copland's Dictionary.

ment of the constitutional forms of syphilis, at an early period of the history of that affection, when it assumed a pestilential form.

The results obtained by chlorine and iodine were to a certain extent successful; and the same may be said of the inhaling of the fumes of belladonna and stramonium, of turpentine, of tar, urged by Crichton and Sir John Forbes, and of the vapors of muriate of ammonia—all of which are still recommended by men of eminence in our midst, and are to this day resorted to, particularly in asthma and in bronchial affections, while turpentine has been warmly eulogized by Skoda, in gangrene of the lungs. Little if any good followed the use of inhalations in consumption, and it was perhaps from being so generally disappointed in their action in this complaint that the profession allowed inhalations to be in the hands of quacks, who, pandering to the popular feeling that remedies to affect the lungs ought to be addressed to them, availed themselves of these agents to allure and to deceive. This was strikingly shown by the use the versatile charlatan St. John Long made of inhalations, which, conjoined to his liniment, formed that treatment through which he became the pet of the fashionable circles in London, and particularly of the female portion. In the room of the handsome adventurer were two enormous inhalers, placed in the interior of a large mahogany case resembling an upright piano. From it flexible tubes ran in all directions, at which numbers of persons were eagerly drawing, while dozens of excited women, of all ages, were waiting until a mouth-piece should be at liberty.* And all over the world were men who, more or less closely, and according to their powers, copied the doings of this ignorant pretender, and have continued until now to do so.

The kind of inhaler which was mostly made use of by the profession for the inhalation of some of the articles mentioned, as of tar and turpentine, and even at times of iodine and chlorine, consisted of the simple instrument still employed for the purpose: a wide-mouthed bottle, with its cork perforated for two glass tubes; the one passing below the level of the medicated fluid, the other, or the one through which the patient inhales, immediately above the level. This form of apparatus may yet be very serviceably resorted to for the inhalation of turpentine, of tar vapor, and of carbolic acid. An apparatus, suitable alike for the inhalation of moist and dry vapors, and arranged to insure their ready disengagement, was suggested by Dr. Snow;† and a similar apparatus has been recently much employed by Mandl in treating chronic bronchitis.‡

Such, then, is an outline of the history of inhalations

---

* A Book about Doctors, by Jaeffreson. 1860. Chapter "St. John Long."
† London Journal of Medicine, vol. iii.
‡ See Stillé's Therapeutics, vol. i. p. 64.

by vapors and gases. But there were some practition-
ers of good repute who endeavored to bring solids and
fluids in contact with the diseased membranes. This,
too, or at all events the use of powders by insufflation,
had its origin among the physicians of antiquity,
though the most systematic attempts to accomplish
this purpose belong to the present day, and insuffla-
tions of alum, of nitrate of silver, and other articles,
through tubes so curved that they could be passed as
far as the larynx, have been frequently tried. These
attempts were greatly stimulated by finding that, in
those whose occupations expose them to it, minute
particles traverse the bronchial tubes and become im-
bedded in the lungs, as has been proved to be the case
in the phthisis of coal miners, and knife-grinders, and
millstone makers.* Moreover, we know that those

---

* See particularly a case by Peacock, in Brit. and For. Med.-
Chir. Review, Jan. 1860; and on the coal miner's lung, Green-
how, in Transact. of Path. Society of London, 1865. Elaborate
papers on the subject have just appeared by Zenker, and by
Seltmann in Deutch Archiv für Klin. Med., quoted in Schmidt's
Jahrb. No. 11, 1866.

Quite recently my attention has also been called by Dr. Flem-
ing, a very intelligent physician of Pittsburg, and formerly a
resident physician in the Pennsylvania Hopital, to the rapidity
with which men repairing copper-work which has been used
for steam, water, etc. become affected by the impalpable oxide
of copper therein formed. He adds in his letter: "Exposed
myself for a few minutes to the same influence, and the quantity breathed exceedingly minute, I was astonished at the
great effect I experienced: constriction about the chest, most
active salivation, etc."

who talk most while exposed are most subject to pulmonary complaints; for among gunsmiths and workers in steel we are told that the ones who suffer most are "les ouvriers bavards et ceux qui ont l'habitude de chanter en travaillant."* But, notwithstanding these proofs that the lungs can be reached by pulverized substances, and the zeal with which the subject of insufflation of powders has been pursued by several observers, it cannot be said to have shown itself of much use therapeutically. Indeed, its only demonstrable value has been in cases of laryngeal disease.

The same may be said of the application of fluids to the disordered mucous membranes, as was so constantly and so skillfully done by Dr. Horace Green. Serviceable beyond all doubt in affections of the larynx, their injection into the bronchial tubes has not a very wide range of utility. For, besides the great difficulty of accomplishment, the diseases to which this mode of practice is suited are not many, and the certainty that the liquid reaches the really affected parts is not great.

Thus, then, neither vapors and gases, nor solids, nor fluids yielded results that could be looked upon as encouraging with regard to the topical treatment of diseases of the organs of respiration. Nay, if we except the inhalation of the vapors of tar and turpentine, and a few others above mentioned, the whole sub-

---

* Guérard. Comptes Rendus de la Société d'Hydrologie Médicale.

ject was receiving very little attention from the profession, until a plan of breaking up fluid into very fine particles was proposed by Sales Girons. This has reopened the whole question in how far a local treatment of the disorders alluded to is beneficial or likely to succeed. But, whatever be the verdict on this point— and it is one of the chief objects of this essay to aid in contributing toward that verdict—it is certain that in the formation of a fine spray, or "pulverization of fluids," or "nebulization," or "atomization," we have gained a therapeutic means of value, which has an applicability much wider than merely to the treatment of the respiratory maladies, and which henceforth will be employed, though it be rejected for the purpose for which it was originally intended.

The first experiments of Sales Girons were crude. They were carried on in a room set apart at a watering-place. The mineral water, either in its natural state or impregnated with various medicinal substances, was projected through a tube with great force by means of an air pump placed in an adjacent apartment. As the stream filled the tube the fluid was forced out of six or more capillary openings, and, impinging against the surface of a metallic disk, was broken up into a mist, which the patients—for several inhaled at the same time—breathed. Though protected by appropriate garments, no sick person relished much the dampness and inconvenience of the whole procedure—one, it may be mentioned, in passing, very similar to the previous attempt of Auphan at Lamotte-les-Bains, and differing chiefly in the fluid not being dashed directly against the walls of the apartment where it was pulverized, much in the way as a waterfall which strikes against the rocks is broken up into spray. Unless, then, something more convenient could be discovered—some apparatus which was portable, easily managed, and yet atomized fluid very finely—the experiments would not, in their practical application, have been of much value. After repeated trials, this was successfully accomplished by Sales Girons, and others have since followed in his footsteps. I shall briefly describe these instruments, of which now many forms exist, particularly such as I have found to be of most service, pointing out what I believe to be the chief merits and defects of each.

The first of the portable kind, projected by Sales Girons and made by Charrière, consists of a vessel in which the fluid to be atomized is poured, and which is attached to an air pump placed above it. The air, compressed on the surface of the fluid, drives this through a very fine opening, arranged with a stop-cock, against a small metallic disk, where very minute spray is formed. The condensed fluid passes off through a gutta-percha tube. The amount of pressure is indicated by a manometer. A pressure of from three to five atmospheres is sufficient. A modification of this apparatus, in which the vessel is made of glass instead

of metal, is generally known as the second model of
Sales Girons. But Sales Girons himself has recently
greatly simplified his whole apparatus, left out the

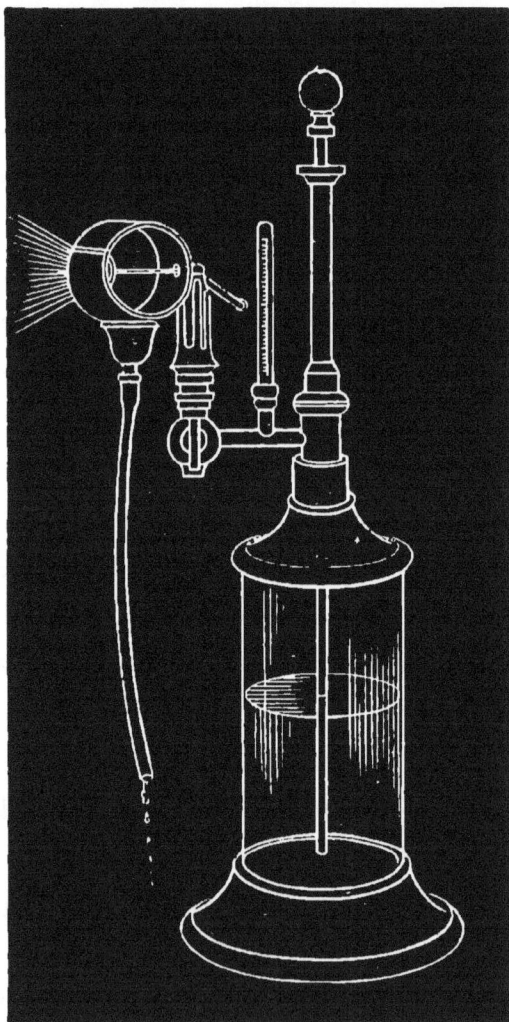

Fig. 1.—The Original Apparatus of Sales Girons.

manometer, and made the instrument lighter and very
much easier of employ. It consists of a pump which

forces the liquid through a fine opening in an inge-
niously arranged stop-cock against a metallic disk, as
in the other nebulizers. The opening in the stop-cock

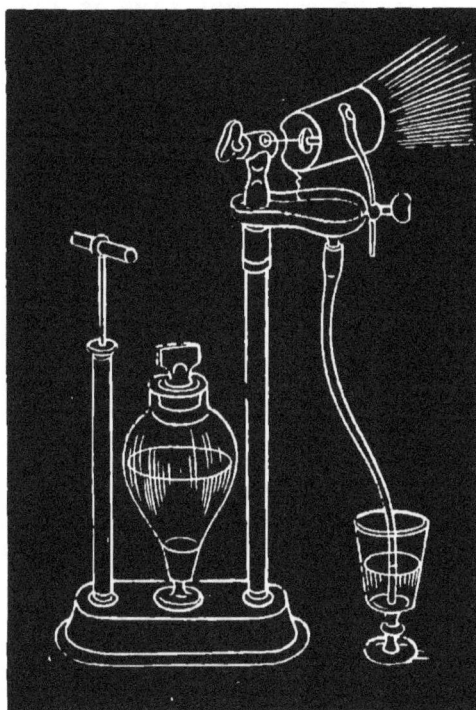

Fig. 2.—The New Atomizer of Sales Girons.

can be increased or diminished in size by simply turning it in a backward or forward direction, and a very
fine spray is undoubtedly obtained by this instrument.

The first-mentioned model of Sales Girons is now but
little used; but it has an historical value, since on the
same principles many others have been constructed—
for instance, the atomizers of Fournié, Waldenburg,
and Lewin. The latter instrument has a very great

advantage in its consisting chiefly of glass: the tube
through which the medicated fluid passes is of glass,
the disk against which the stream of fluid strikes in the
cylinder of glass is gilt, and thus is unquestionably
avoided one of the most serious objections to the atomizers of Girons, and which renders them unfit for the use
of chloride of iron, nitrate of silver, and other articles
that act on metal.

On a similar principle as the last model of Sales
Girons is a nebulizer, in which the pump works by a

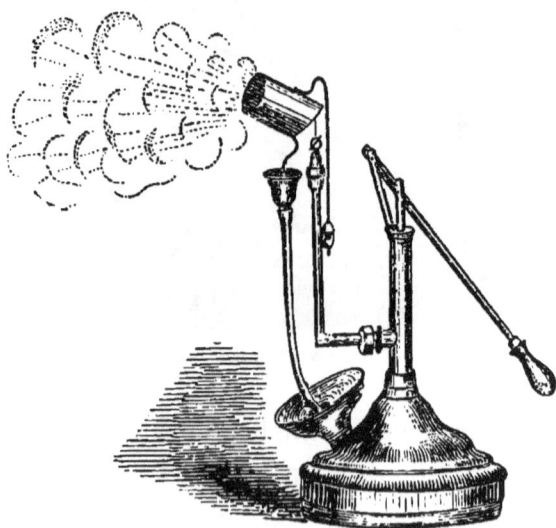

FIG. 3.—The Pump Nebulizer.

long handle, and the fine stream impinges against the
side of the cylinder, but not against a metallic disk.   I
think this apparatus was originally made by Matthieu;
and to it, as well to the last invented one of Sales

Girons, may be attached a little tube, through which a capillary jet passes with such force that it penetrates the skin, and may be used to inject medicines hypodermically.

In all the instruments alluded to the fluid is forced by strong pressure against some firm body, where it is broken up into very fine particles. But another principle has been made use of to accomplish this purpose, namely, the action of a current of air compressed in a large ball, and which, intermingling with the fluid, changes this into a minute spray as it rushes out of a capillary opening. This instrument, the "Nephogène" of Matthieu, is handled with readiness; but it is apt to get out of order, and the spray is thrown with greater force than is usually advisable; moreover, a large quantity of atmospheric air is projected into the air-passages with it. A far better application of using a current of air as the means of atomizing the medicated liquid was made by Dr. Bergson, by employing the same kind of tubes as are now so extensively sold as odorators, for the pulverization of different scents. Two glass tubes, with capillary openings, are placed at right angles to each other, in such a manner that the end of the vertical tube is very close to and about opposite the centre of the capillary opening in the horizontal tube. Through this the air is blown while the vertical tube is dipped in the fluid to be atomized. The air in the latter tube is rarefied; the liquid rises to the capillary open-

ing and is there pulverized by the current of air from
the vertical tube. Two tubes properly arranged are
then all that is strictly required for this simple apparatus, for the air may be blown by the mouth through
the horizontal tube. But in point of practice this procedure is both irksome and unpleasant, and to avoid it
an ordinary Davidson's rubber syringe may be attached
to the horizontal tube; or, better still, a continuous

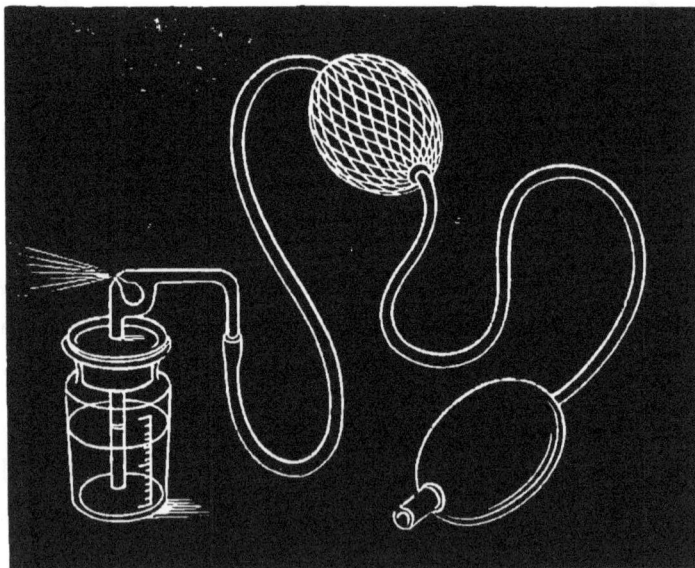

FIG. 4.—The Hand-ball Atomizer.

stream may be obtained, as proposed by Bergson, by a
bellows connected with an air chamber. The bellows
is worked by the foot. Yet more convenient is a similar
arrangement of Andrew Clark, consisting of two balls,
the lower of which is pressed by the hand, and the
upper of which, surrounded by a silk network, acts as
an air chamber.

The principle of Bergson is distinguished by great simplicity, and is very readily applied. The tubes may be made of silver or of glass. Those of the latter are generally preferable. They are much more easily kept clean, and can be used with articles which corrode the former. On the other hand they are more easily broken. Tubes of glass can be kept clean by soaking them occasionally in muriatic acid and water, or pulverizing a little of this mixture through them. If the tubes become clogged, a bristle or a very delicate metallic wire is the best means of removing the obstruction. A pin or a needle is apt to break the fine points. The tubes may be made of any length or calibre. To produce a delicate spray, the openings at the ends, where placed in juxtaposition, ought to be very small. The hole in the horizontal tube may be somewhat larger than the

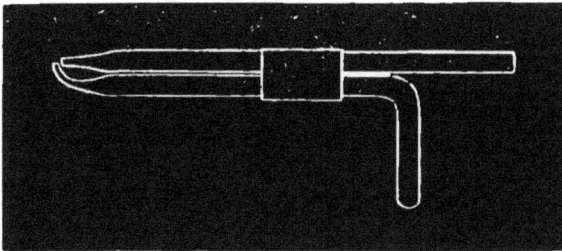

Fig. 5.—Bergson's Tubes of modified shape, united by an India-rubber band.

capillary opening in the vertical tube. A modification in the shape of the tubes, as seen in Fig. 5, was proposed by Prof. Winterich, and is often of great service. By this arrangement we can generate the

spray within various parts of the body. I have had
tubes of the kind made of all sizes and of different
curves—to pass up the nostril, as in the treatment of
catarrh; to place in the ear and reach the membrane
of the tympanum; to apply near the back of the throat,
or immediately over the entrance of the larynx—thus
furnishing a far better means for local treatment than
the ordinary sponge probang, and even better than the
laryngeal fluid pulverizer of Gibb. If it were judged
expedient as a therapeutic means, they could be so
shaped as to throw a spray even into the interior of the
uterus or bladder.

Similar in some respects to the arrangement just
mentioned is Maunder's spray producer. The an-

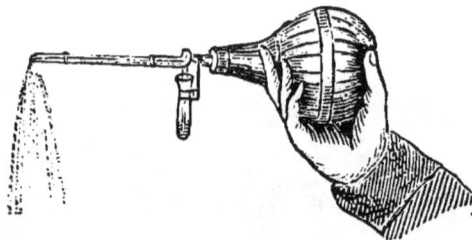

FIG. 6.—Maunder's Laryngeal Spray Producer.

nexed wood-cut shows that it consists of an India-
rubber bottle perforated at the base so as to admit,
when emptied, of instant refilling with air; of an
upper or air tube communicating at one end with the
bottle and terminating at the other in a capillary opening; of an under or medicine tube dipping at one end

into a medicine glass and ending at the other in a capillary orifice.

Bergson's tubes are also employed in an atomizer, invented by Dr. Oliver, of Boston; though here the spray is still further broken up and converted into a fine mist by impinging against the walls of the glass vessel in which the tubes are contained. At the same

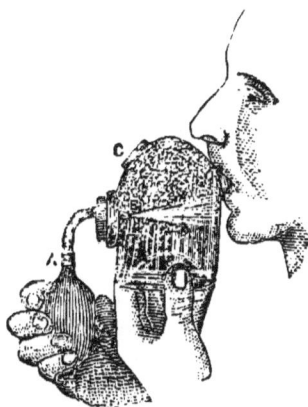

Fig. 7.—Oliver's Atomizer, as made by Codman & Shurtleff, Boston.

time the face is by this arrangement completely protected. For the use of caustic solutions, and where it is desirable to atomize only small quantities of the fluid, this form of atomizer claims several advantages.

The instrument arranged by Richardson for local anæsthesia is a modification of the hand-ball atomizer. The principle of Bergson is employed, with the addition of that of pressure caused by air forced at the same time into the bottle by the air bag. The air and the liquid are jetted out together from the orifice

3*

of exit, and thus the spray is produced. With ether of specific gravity of ·720, or less, the part may be frozen by working the apparatus about one minute, and with rhigolene—the fluid introduced by Dr. Bigelow—in from 5 to 15 seconds. The pain produced by rhigo- lene is, I am informed by Dr. Keen, who has made a number of comparative experiments on the subject, very much less than that caused when ether is em- ployed; which, in truth, occasions often almost insup- portable suffering. Rhigolene, it may be added, is not used with Richardson's tube, but with metallic tubes on the Bergson principle; and the apparatus for either agent may also be made available as an ordinary spray producer in the treatment of throat affections.

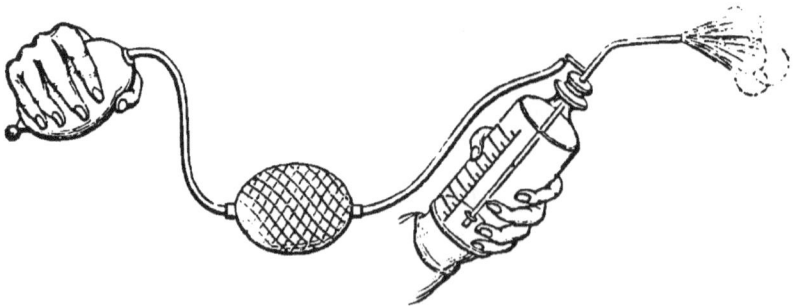

Fig. 8.—Richardson's Spray Producer.

But to return to atomizers proper: we have dis- cussed those in which air acts as the forcing power, or in which the fluid is driven by a piston through a narrow opening. They require, of course, a certain amount of exertion on the part of the patient, or an assistant. This is obviated in the ingenious apparatus of Siegle,

who substituted steam as the motive power.   Adopt-
ing the arrangement of tubes of Bergson, he has added

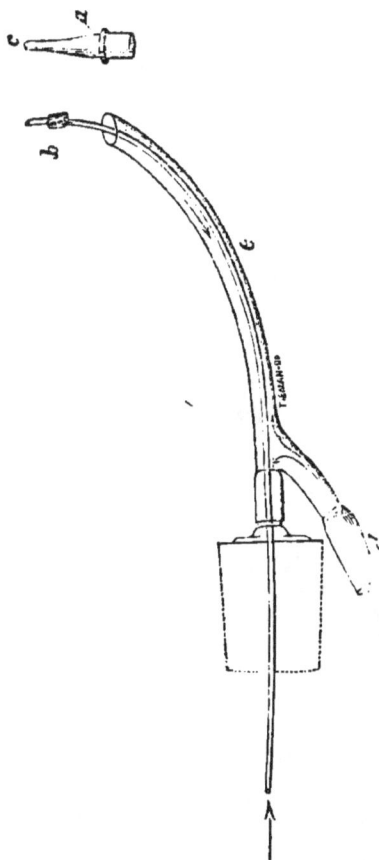

FIG. 9. — Wood-cut illustra-
tive of the interior arrangement of Richardson's spray
producer. a, adjustable con-
ical cap; e, hollow curved tube
containing within capillary
tube, and this near its upper
extremity perforates cylinder
of metal b; d, tube to be attached to the India-rubber
tubing. The tubes in this apparatus, as made by Tiemann
& Co., N. Y., are of silver.

a small boiler, made of metal or of glass, in which
steam is generated by means of a spirit lamp.   The
steam plays the part of the compressed air, and as it
escapes projects as a fine spray the liquid placed in the
cup.   The degree of pressure is indicated by a thermo-
barometer, marked 1 and 2.   It is safe enough to let
the mercury range between 1 and 2; above this there

is some danger. By lowering the flame the steam is
generated with less rapidity and force. A small lamp
under the glass cup containing the medicated fluid heats
this, and thus the inhalation may, if necessary—which,
however, it extremely rarely is—be given very warm.
Fig. 10 represents Siegle's apparatus in its most perfect

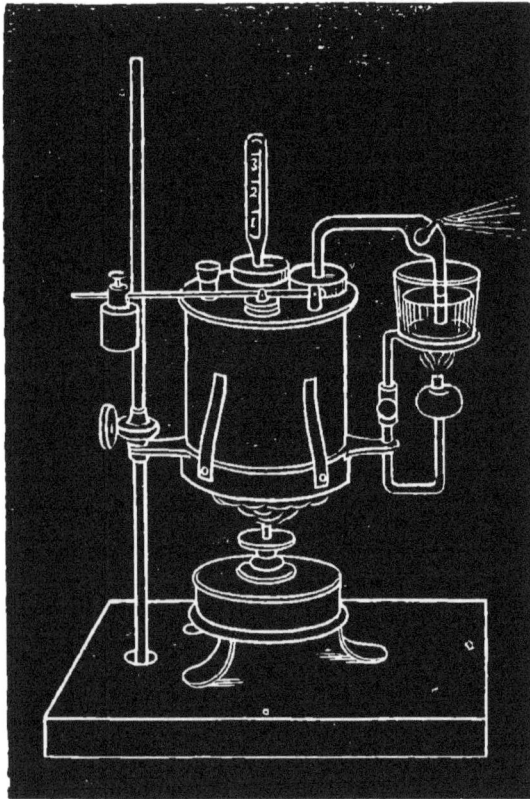

FIG. 10.—Siegle's Large Atomizer.

form. In addition to the thermo-barometer we find a
safety-valve.

Now, on Siegle's principle, numerous instruments
have been constructed. The size of the apparatus, its

shape, the lamp, have been modified; but few of the modifications are really improvements. It may, however, for ordinary purposes, unquestionably be much simplified; and Mr. Gemrig, an instrument maker of Philadelphia, has constructed, according to a design I gave him, a steam atomizer, which is both simple and very convenient. It consists of a copper boiler, with a spring safety-valve in place of the thermo-barometer. By unscrewing the safety-valve the water can be poured into the boiler. This fits into a metallic tube, at the

Fig. 11.—Siegle's Apparatus, with water-gauge (B) and valve (F), as modified by Krohne, and made by Otto & Reynders, N. Y.

bottom of which a spirit lamp is placed, the flame of which can be heightened or lowered. The atomizing tubes are inserted into a cork, or a perforated piece of gutta-percha, which is readily fastened by a metallic rim

with a bayonet catch. In some of Siegle's instruments
this point is omitted, and the cork or piece of rubber is
liable to be blown out when the steam is generated.
In his largest apparatus (Fig. 10), screws hold down
the India-rubber at two of the openings. Leaving
out the thermo-barometer is, in any apparatus, excepting those for purposes of very accurate study, a great
gain. It is irksome to the patient to be constantly
watching it while inhaling, and is apt to get broken.

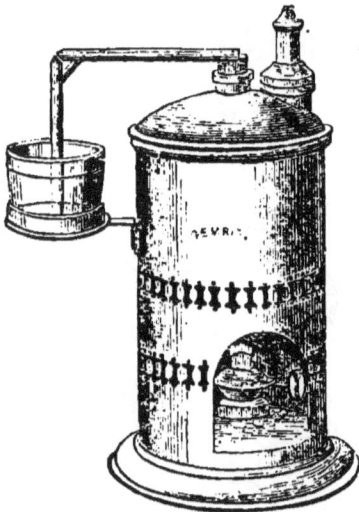

FIG. 13.—Boiler with Tubes attached
and spring safety-valve.

FIG. 12.—Simple form of Steam
Atomizer.

The simple apparatus just
described is shown in Fig.
12; its component portions
are seen in Figs. 13, 14, 15,
and 16. When in action,
the boiler should be about two thirds full of water,
and after the medicated fluid has been pulverized, the
tubes should be cleansed by letting the instrument
nebulize pure water.

As the various apparatuses for atomizing liquids have
been passed in review,—and in so doing I have en-

deavored to describe rather the different principles that have been suggested for their construction than attempted to give a complete list of all the instruments

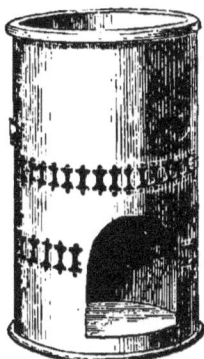

FIG. 14.—Metallic ti be at the up｜er p｜ rt of which the boiler is placed and at the lower part the lamp.

FIG. 15. — Lamp, with screw to raise or depress the wick.

FIG. 16. — Cup for medicine, and rest with slide to hold it.

which have been proposed,—I may now state what I believe to be their relative efficiency. The most perfect as well as the most efficient is that of Siegle, or some of those framed on his plan. The steady stream, the possibility of working the apparatus without fatigue and without an assistant, the small quantity of medicated fluid required—while in most of those which are set in action by a pump the quantity is very great—are all features of pre-eminent value. Then the minute division of the fluid and the warmth of the spray that reaches the respiratory organs are also points of decided consequence. But under some circumstances other atomizers will answer better; for instance, in the treatment of affections of the throat or nares, or even in many of those of the larynx, the nebulizer depicted Fig. 3, or Maunder's, or Bergson's tubes with the bellows attached, may, for the most part, be more easily and advantageously employed. Again, this apparatus may be resorted to when the patient is in the recumbent position; when he lies in such a manner that an instrument with a lighted lamp cannot with safety be placed near him; when the heat of the weather renders it annoying to use a steam apparatus; when a cold rather than a warm inhalation is therapeutically desirable; when, on account of the time lost, it is inconvenient to have to generate steam; or when the patient is careless, or not intelligent enough to learn to use an apparatus which requires both care and some intelligence to use properly. Then, in very many cases, particularly those of affections of the fauces and windpipe, the quantity of medicated fluid necessary is but small, as, for example, when solutions of caustic are required. And here the hand-ball atomizer or some of its modifications will be found very convenient. Indeed, I think that, excepting the steam nebulizer, it is the best form of spray producer—superior to that of Sales Girons, and to the one depicted in Fig. 3; and for the purposes just alluded to it is even better than the steam atomizer, while its ready use and portability, the fact that the fluid can be injected without any or with but slight co-

operation on the part of the patient—in children especially a very great advantage—and the manifold employ
to which it can be put, are additional recommendations.
Yet it cannot be said to take the place of the steam
atomizer in the treatment of diseases of the lung. It
becomes too irksome and fatiguing to work it long
enough to produce the quantity of spray required for
each inhalation.

4

CHAPTER II.

THE MODE OF ADMINISTERING INHALATIONS.

BUT whatever apparatus be employed, there are some points to be observed in the use of inhalations, the neglect of which will seriously interfere with the benefit to be derived from them, and may, indeed, cause them to be abandoned in disgust. And, in the first place, we should instruct our patient in the application of the instrument, show him how to keep it clean, how to tell when it is working properly. This implies that he should have one in his possession. In truth, excepting when resorted to in certain affections of the larynx and fauces, or for a mere temporary purpose, and particularly when required in chronic diseases of the lungs, the inhalations ought not to take place at the physician's office. The patient must employ them once or several times daily, and unless he can attend to them himself, in the same manner as he knows how to take his tonic pill or his cough mixture, the treatment will be inefficient. It is evident, then, that he must, as a rule, carry on the treatment at his own house. At the first inhalation the physician ought always to be present.

When the patient is ready for the inhalation, he should sit in front of the apparatus, in a convenient position, and in such a manner that the spray is formed on a level with his mouth. The mouth must be kept wide open and the head be slightly inclined backward. The distance proper to sit from the spray-producing tubes varies. When he begins the inhalation, he ought to be about six inches from them. This distance may be increased from one to two feet, according to the object we have in view. If we wish the patient to inhale a large quantity of the spray, and at a comparatively high temperature, let his mouth be near to the apparatus. And we direct him to take deep breaths when we desire to reach the bronchial tubes and more distant portions of the respiratory channels, while we insist upon shallow breathing if our intention be to act on the fauces and upper part of the air-passages. But under no circumstances should he breathe in such a manner as to be fatiguing to him; and it will be often necessary to restrain him from respiring with much effort and hurriedly.

In persons with sensitive mucous membranes the act of inhalation causes considerable cough. But even this can ordinarily be avoided by letting them breathe at first the spray warm and close to its point of production, and by commencing with inhalations of pure water. In most cases, after a few inhalations, no cough is produced; nay, strange to say, coughing is more apt to occur when the inhalation is over than while in progress.

A point always to be attended to is to see that the tongue is not in the way of the current, and that the spray can really reach the back of the throat. It is well to direct the patient to press his tongue against

FIG. 17.—Face Shield.

the floor of the mouth. If he cannot easily do this, a tongue depressor may be employed; but I have found it more advantageous to insert a short small glass speculum, from three to four inches in length, into the mouth. This keeps the tongue out of the way, and yet he can breathe very readily through the tube. To prevent the face from becoming wet, a face shield may be employed, preferably made of glass or wood. It can be held by the patient, or attached to the atomizer, as it is in the very serviceable instruments made by Dr. William Read, of Boston; or fixed near to it in

shape of a screen, as suggested by Dr. Beigel; or
placed on a movable supporter in front of the apparatus, as is done by Codman & Shurtleff, of Boston, in
the excellent, safe, and durable steam atomizer they

Fig. 18.—Steam Atomizer of Codman & Shurtleff, with face shield.

make. In any case the shield can be so arranged with
a depressed rim, into which a gutta-percha tube is
·fixed, that the drops of fluid which collect flow into a
glass. But I find that patients often prefer dispensing
with a face shield. It is, of course, always necessary
to protect the clothes with a napkin or towel, and to
have a vessel at hand into which any of the fluid which
may accumulate in the mouth can be expectorated.

As regards the frequency and the time of the inhalations, it is difficult to lay down general directions. But
this much is certain: for the treatment to be effective,
the patient should inhale daily, and breathe the medicated spray for about ten minutes or longer, taking in

that time about one hundred respirations or more, and resting for some seconds after inhaling continuously for a few minutes. In many cases it is better to have him inhale twice or three times daily; and it is always, perhaps, more convenient to let him inhale a certain quantity, say one ounce of the medicated fluid, than to annoy him by directing him to count the frequency of the act of breathing. With a well-constructed steam atomizer, the time of each inhalation should be from ten to fifteen minutes, and about the same time is required at each sitting with the nebulizer of Sales Girons, or any of those working with a pump. The first inhalations ought always to be short, so as to accustom the patient to them; and it is astonishing how, though they irritate him at the beginning of the treatment, he becomes less and less sensitive to them. The patient ought never to inhale on a full stomach, and should abstain from going out of doors for a quarter of an hour after the inhalation.

In these remarks I have had chiefly reference to the treatment of pulmonary affections, and secondarily to that of laryngeal diseases. But as the form of therapeutics under consideration applies also to disorders of the fauces and adjacent structures, I may briefly indicate in what manner the mode of procedure is to be modified in their treatment. The inhalation is of shorter duration, and need not, excepting in certain urgent cases, be done so frequently. The greatest care should be exercised to cause it to reach the affected spot; and to effect this I have often found the method above mentioned, by passing the current through a small glass speculum introduced into the mouth, very efficient. It has, moreover, the advantage of bringing the diseased surface thoroughly into view, and of limiting the action of the pulverized fluid much more completely to it. When the spray is thus passed through a speculum, and even, if the tongue be not in the way, by simply directing the jet to the affected textures, we can obtain results which are not attainable by means of ordinary local applications. The spray reaches parts more readily than the probang, and in certain cases produces a preferable and more permanent action. Its superiority over gargles is manifest, touching structures never or but scantily reached by these, as for instance the posterior wall of the pharynx; so striking is this superiority, that for really useful purposes the day of gargles has passed. Even in enlargement of the tonsils, I have used pulverized fluids with considerable success. In one case, particularly, that of a little boy greatly troubled with chronic pharyngitis and enlarged tonsils, and liable on any exposure to acute exacerbations, they effected a cure which no other means could have accomplished. Each attempt to reduce by caustic or astringent solutions the tonsils, which nearly blocked up the half arches, was a signal for an outburst of passion and for violent resistance on the part of the child. To the treatment by pulverized fluids, such as of strong solutions of tannin, he submitted without objections.

Let me add, in concluding the mode of inhalation and the modification necessary to adapt it to individual parts, that the force of the current is an element also to be taken into account. A very strong current is not suitable for pulmonary affections; it is more suitable for those of the fauces. The spray projected with too much force is apt not to enter the air-passages, but to condense on the walls of the throat.

CHAPTER III.

THE PENETRABILITY OF ATOMIZED FLUIDS INTO THE AIR-PASSAGES.

IN the preceding considerations, and while discussing the mode of applying the medicated spray, it has been assumed that this can be made to reach not only the fauces and pharynx, but the respiratory channels. Now, with reference to the former point there can be no question; but much skepticism has prevailed as regards the entrance into the air tubes of the pulverized fluids; and before passing on to indicate the uses of these agents, it is incumbent to inquire into the evidence on which it has been assumed that they penetrate into the lungs, or even into the larynx. Not to mention further the well-known facts alluded to in connection with the coal miner's lung, and which have a strong bearing on the subject, the evidence of fine particles reaching the air-passages is of two kinds: first, that furnished by demonstrative experiments made on animals and man; and, secondly, that attained by perceiving the effects of inhaling the atomized liquid, particularly its prompt effect in producing or allaying spasm, in checking hemorrhage and the like.

To the first category belong the interesting experiments of Demarquay, which, repeated in the presence
of Poggiale, were used by him as the basis of his able
report to the French Academy. Dogs and rabbits,
with their mouths forced open and their nostrils closed,
were made to inhale for five minutes a pulverized solution of perchloride of iron, of the strength of 1 part of
the iron to 100 of distilled water. The animals were
afterward killed, and throughout the larynx, trachea,
bronchial tubes, and even in the lung structure, the
presence of the persalt of iron was clearly detected by
the production of Prussian blue with the ferrocyanide
of potassium and acetic acid. But it was objected to
these experiments that they were made on animals,
and that all the circumstances were such that it was
unfair to infer that the same results could be obtained
in man. To meet these clamors, Demarquay experimented on a nurse at the Hospital Beaujon,
who breathed through a canula in her trachea. She
inhaled a pulverized solution of tannin while the
tracheal opening was closed with a slip of paper,
moistened with a solution of perchloride of iron, and
held in place by means of a strip of sticking-plaster
and a napkin. After inhaling for a minute, the sticking-plaster and the paper were removed, and a piece
of paper impregnated with a solution of perchloride
of iron was pressed, with the aid of a delicate forceps,
into the trachea. In the first two experiments there

was no reaction; in the third, the black discoloration of the paper proved that the solution of tannin had entered the air-passages. Fieber,* in Vienna, repeated the experiment on a man twenty-two years of age. It only succeeded on the fourth attempt. A few months since I tried it three times on a man about twenty-six years of age, a patient under my charge at the Pennsylvania Hospital. He had cut his throat in an attack of mania a potu three months before he came under my observation, and wore a canula. I made him, after removing the canula, and while closing the tracheal opening with the finger, as done by Demarquay in his third, the successful experiment, inhale, for nearly two minutes, and by means of a steam atomizer, compound solution of iodine, 15 drops to the ounce, and then passed a thoroughly starched linen rag into the opening. No satisfactory reaction was perceived; neither was it when a canula, covered on its upper surface with a starched piece of linen, was employed. Nor was an experiment with tannin and chloride of iron more successful. A laryngoscopic examination showed the cause of failure. The false vocal cords were tumid, greatly engorged, rigid; and the motion of the cords was hardly perceptible even when the man, while the tracheal opening was closed, was doing his best to breathe through the mouth. They were

---

* Die Inhalation, etc., 1865.

scarcely dilatable, though after a course of laryngeal catheterization they became so.

The difficult success of Demarquay and Fieber, the total want of success of Fournié, who experimented on the case of Demarquay, were chiefly due to the great obstacles in completely closing (owing to the anatomical relations of the parts) the tracheal opening—an indispensable condition for preventing the experiment from being a vain one; and to the inability of the patient to remain but for a very brief period deprived of the canula. In this respect my case was better suited, yet it presented other and greater hinderances. In truth, a union of all conditions favorable to complete success is rare; hence, while we may claim the positive results obtained as conclusive, the failure to obtain these results, it is evident, is not equally so.

Still, though we can very rarely on man demonstrate to the eye the passage of atomized fluids down the trachea, it is easy to satisfy ourselves of their entrance into the larynx. This observation which I have made will prove it. Let a person inhale pulverized distilled water, to which some drops of a solution of blue or red aniline have been added. Let him then be immediately examined with a laryngoscope; an intense color, visible on the vocal cords and at the beginning of the trachea, will show where the aniline has reached. Bataille,* inhaling a solution of rhatany, noticed on

* Gazette Hebdom., 1862.

himself, by means of the laryngoscope, the red discoloration of the larynx and trachea. He also expectorated for a whole day subsequently a reddish sputum, which, from its character, he believes to have proceeded from the bronchial tubes.

Yet another demonstration of the penetration of the atomized liquid is furnished by post-mortem inspection of pathological processes. A number of the rabbits Demarquay experimented on, and which were not further interfered with after they were made to breathe the solution of the perchloride of iron, died of pneumonia, generally of very circumscribed kind. Trousseau reports the same consequence from inhalations of tannin in a lady, who, finding herself benefited by them, inhaled for several hours daily until a fatal pulmonary inflammation supervened. Still more significant are the cases reported by Lewin and Zdekauer. In Lewin's case, inhalations of chloride of iron were used to arrest a hemorrhage from the lungs. The patient soon afterward died, and little particles of iron were found by Schulz, the chemical assistant of Frerichs, in a cavity in the upper lobe of the right lung.* In the similar case of Zdekauer,† the Russian professor, assisted by Holm, detected a far larger quantity of iron everywhere in the tissue of the lung than appertains to the blood it contains.

---

* Inhalations Therapie, p. 190.
† Quoted in Wiener Med. Wochenschrift, 1861, No. 30.

These facts, some of which were elicited and dwelt
upon in the report and discussion before the French
Academy, and which convinced, as we may judge by
the adoption of all the points of the report, that critical
body of the penetrability of atomized fluid, can leave,
then, no doubt on the subject. But there is another
matter which it seems to me may be advanced as evidence, namely, the immediate effect perceived from the
inhalation of certain articles. For instance, if cold
water be inhaled, a sensation of cold in the larynx and
trachea, extending thence into the chest, is very common. Many articles in strong solution, and tannin
may be mentioned among them, give rise to a feeling
of oppression and of violent burning in the chest.
From the inhalation of a strong solution of alum, I
have seen on two occasions asthmatic, wheezing
breathing very speedily produced, with loud dry rales
discernible at various parts of the chest; the attack
lasted for eight or ten minutes. Further may be mentioned what, in a patient under my care, affected with
bronchorrhœa, I have repeatedly observed, the sense of
tightness in the chest attended with a very greatly, I
might say immensely, diminished expectoration very
soon after inhaling any strong astringent solution, particularly of alum or of tannin.

The question of the pulverized fluids reaching the
respiratory channels is thus, to any unprejudiced mind,
no longer one of unbelief. But it still remains to be

solved how much of a given solution arrives there, and
what proportion finds its way into the deeper textures.
Again, is the temperature a modifying agent; do chemical changes take place in the atomized liquid when inhaled; and under what circumstances is the passage of
the spray prevented? To enter into these questions at
any length would necessitate long and tedious discussion, involving allusion to many chemical and physical
laws. Let me merely state that, though attempted to
be solved by Waldenburg, with great care, we do not
know how much of the fluid gets into the respiratory
structures; and that we shall have to decide its passing
in any quantity chiefly by physiological and therapeutic
experiments. Chemical reactions only take place in
certain articles pulverized, as in sulphurous waters.
The temperature of the stream varies with the temperature of the surrounding atmosphere, the apparatus
employed, the distance of the patient from the spray-
producing tubes, and the temperature of the fluid to be
pulverized. Bearing this in mind, we can ordinarily
regulate the heat of the stream without much difficulty;
and in point of fact it must always be recollected that
it takes very readily and speedily the temperature of
the surrounding air. Practically, therefore, the subject
of the temperature of the spray does not occasion much
perplexity. The spray from a steam atomizer is
warmer, particularly when inhaled rather near to its
point of production, and from its comparative warmth
is generally more acceptable to the person inhaling.

With regard to the circumstances interfering or preventing the passage of the spray into the respiratory
passages, I have already alluded to the intensity of the
current. I may add that breathing through the nose;
the tongue not being sufficiently depressed; the head
being inclined forward; and all other postures which
would change the angles favorable to the progress of
the pulverized fluid, or interfere with the freedom of
respiration, are obstructing elements. I have tested
these points experimentally by letting a man breathe
an atomized solution of aniline in conditions unfavorable to the entrance of the spray, and have then examined him with the laryngoscope to obtain a view of
the discolored laryngeal membrane. I found that
though he may be kept inhaling steadily for four or
five minutes, there is scarcely a perceptible alteration
in color at the beginning of the respiratory passage,
and it is therefore highly improbable that any of the
pulverized liquid should have passed lower down.
Many of the negative experiments, whether on man or
on animals, have been, I think, clearly due to a neglect
of the points mentioned.

It was necessary to discuss these hindering causes,
because it was necessary to indicate in how far we
could guard against them, and while showing that
they might explain some of the discrepancies of different observers, to make evident their bearing in estimating the effects of medicines employed by inhalation.

CHAPTER IV.

DOSES OF MEDICINES FOR INHALATION.

THE question of doses is one far from easy to determine, and can only be fixed experimentally—can, in other words, only be settled by a careful study of respiratory therapeutics, in which due importance is attached to the sources of fallacy already indicated, and in which they are avoided. Moreover, the dose varies with the apparatus; or rather, though the dose be the same, to obtain that dose in the mouth we may have to use more of the medicated fluid or a stronger solution with one atomizer than with the other. Thus, in the apparatus of Bergson, it takes, according to a calculation of Lewin—whose own glass atomizer is, however, still more wasteful of the medicated liquid—eight ounces of fluid, which quantity can be pulverized in from twelve to fourteen minutes, and of which three-eighths only arrive at the mouth, even when the patient is suitably near to the spray-producing tubes; a point of course which also influences the estimate of doses. To insure, therefore, three ounces of spray reaching the oral cavity, we must use an amount of solution nearly three times the dose required. With the hand-ball atomizer

5*

with fine spray tubes, I do not think the quantity
lost is nearly as great; with ordinary care, and placed
rather close to the patient, fully three-fourths reach
the mouth. In Siegle's steam apparatus, one ounce is
atomized in about twelve minutes, and perhaps not more
than a fourth is lost, but the steam dilutes the solution
prepared by nearly one-half; a solution in the cup of
ten grains to the ounce would therefore be reduced by
the vapor to between five and six grains to the ounce,
and of this about four grains would be really inhaled
through the mouth.

Speaking generally, the dose to be given does not
vary materially from that employed internally. But,
with reference to narcotics, this does not hold good, as
they are readily absorbed and act efficiently in smaller
doses. Concerning astringents, too, though they are
often employed in doses approximating in their strength
those for external use, when designed to reach the
deeper structures, we must, bearing in mind the deli-
cacy of these textures, carefully graduate the dose.
Any agent which is soluble in water, or in a watery
infusion of an aromatic, or in very dilute alcohol, can
be used by an atomizer. Substances soluble in glyce-
rin, or capable of being suspended in thin emulsions,
may also be employed, but not, as a rule, satisfactorily.
I now subjoin a table, in which the dose is calculated
to the ounce of water, for any form of steam atomizer
throwing a fine spray. It represents the articles which

have been most employed, and there is scarcely one in the table that I have not used in the doses mentioned. Where the dose is not stated from personal knowledge, or where a particular kind of application is alluded to, I have added the name of the observer. I have also indicated the pathological conditions to which the doses are suitable.

TABLE OF DOSES FOR INHALATION.

ALUM, 10 TO 20 GRAINS.—In this dose suitable to chronic catarrhal affections of pharynx and air tubes, particularly in bronchial affections with excessive secretion, when, as in most inflammatory conditions of the respiratory mucous membrane, it may be advantageously united with opium. In rather larger doses, 30 grains to the ounce, useful in pulmonary hemorrhage. Is, as an astringent, generally more of a sedative and more suited to conditions of irritation than tannin. (Fieber.)

TANNIN, 1 TO 20 GRAINS.—Useful for the same affections as alum. Employed in cases of laryngeal ulceration and excrescences, in œdema of the glottis (Trousseau), in croup. Here, as well as in pulmonary hemorrhages, in large doses. In ordinary cases of laryngeal or bronchial disease, begin with a small dose. If the remedy occasion much heat and dryness, it is not to be employed.

IRON (perchloride of), ⅓ TO 2 GRAINS.—In earlier stages of phthisis. In chronic pharyngitis or laryngitis may be used stronger. As a weak inhalation in hysterical aphonia. Of greatest strength in pulmonary hemorrhage, 2 to 10 grains to the ounce, or 10 to 40 m. of Monsel's salt to the ounce. The lactate, citrate, or phosphate may, in ordinary cases, in which we wish a non-astringent salt of iron, be also used, though they are not, on the whole, as available as the chloride.

NITRATE OF SILVER, 1 TO 10 GRAINS.—In ulcerations of pharynx and larynx, in follicular pharyngitis. A face shield is always to be used. 10 grains to the ounce only in cases of ulceration.

SULPHATE OF ZINC, 1 TO 6 GRAINS.—In bronchial catarrh with excessive secretion. In aphonia, connected with chronic laryngeal catarrh.

CHLORIDE OF SODIUM, 5 TO 20 GRAINS.—Promotes expectoration and diminishes sputa; and employed in phthisis.

CHLORINATED SODA (*Liquor Sodœ Chlorinatœ*), ½ TO 1 DRACHM.—In bronchitis, with offensive and copious expectoration; in phthisis.

CHLORATE OF POTASSA, 10 TO 20 GRAINS.—In chronic catarrhal laryngitis and bronchitis. In subacute or chronic laryngeal and pharyngeal congestion, particularly when attended with a feeling of dryness.

MURIATE OF AMMONIA, 10 TO 20 GRAINS.—In laryngeal and bronchial catarrh, acute as well as chronic. To promote expectoration; also in capillary bronchitis. The dose best borne is not above about 10 grains to the ounce, though as much as two drachms to the ounce have been employed. (Siegle.)

OPIUM (watery extract of), ¼ TO ½ A GRAIN.—In irritative coughs, and as an adjunct to allay irritation. Also for its constitutional effects. Dose of tincture of opium 3 to 10 drops. Acetate of morphia one-twelfth to one-eighth of a grain has been administered, but large doses require much caution.

CONIUM (fluid extract of), 3 TO 8 MINIMS.—Irritative cough; asthma; feeling of irritation in larynx.

HYOSCYAMUS (fluid extract of), 3 TO 10 MINIMS.—Spasmodic coughs; whooping-cough. One-half a grain of the extract, gradually increased, or the tincture may be employed.

CANNABIS INDICA (tincture of), 5 to 10 Minims.—In spasmodic coughs; phthisis.

IODINE (*Liq. Iodinii Compos.*), 2 to 15 Minims.—In chronic bronchitis; in phthisis.

ARSENIC (*Liq. Potass. Arsenit.*), 1 to 20 Minims.— Nervous asthma. (Trousseau.)

TAR-WATER, 1 to 2 Drachms of officinal solution.—In offensive secretions from bronchial tubes; in tuberculosis; as an antiseptic in gangrene of lungs.

TURPENTINE, 1 to 2 Minims.—In chronic bronchitis with offensive secretions; in bronchorrhœa; in gangrene of lungs.

LIME-WATER, used of officinal strength, or stronger.— In diphtheria; in membranous croup.

WATER, Distilled.—Cold, in pulmonary hemorrhage. Warm water in asthma, in croup, in bronchitis.

It is always preferable that the solutions should be made by the addition of distilled water; and it saves much annoyance in the working of the atomizer if some of them—for instance, those of tannin—are strained. In some cases the dose recommended cannot be borne at first. It is, indeed, always best, excepting if the prompt action of a narcotic be needed, to begin with small doses, and educate, as it were, the respiratory mucous membrane to tolerance.

What has been stated applies only to doses for atomization. The quantities for the ordinary inhaler or where gases are inhaled cannot be so accurately fixed. I have used tincture of iodine one drachm to the pint of hot water in cases of chronic bronchitis and

of early phthisis, without seeing, however, any decided
effects from it. Carbolic acid may be employed in the
same strength and several times daily, with, I think,
appreciably fair results in chronic bronchitis and to
further expectoration. It is not necessary, of course,
to change the solution each time the inhalation is resorted to. Pouring boiling water on tar in about equal
proportions, or in cases of susceptible mucous membranes in less strength than this, and using the inhalation twice a day or oftener, ten minutes at a time is
of service for the same purpose, and has been even
recommended in tubercular cases. It was by pouring the essence of turpentine upon boiling water and
directing the patient to inhale the vapor for fifteen
minutes every two hours that Skoda obtained the
recovery of the cases of gangrene of the lungs he
published.

Chlorine and iodine gas cannot be inhaled unless
diluted with air. The inhalation of chlorine recommended by Pancoast, in cases of catarrhal aphonia with
subsequent debility of the cords, was effected with an
ordinary glass retort and a glass funnel, having some
filtering paper at the bottom. In the bowl of the retort
was placed a solution of chloride of sodium or lime,
and in the glass funnel a weak solution of sulphuric
acid in water. As the dilute acid fell drop by drop
into the bowl of the retort, chlorine was very gradually
liberated and breathed from the end of the instrument.

The inhalations were repeated two or three times a day; each lasted a few minutes.*

Various means have been suggested to obtain the fumes of muriate of ammonia, which has been much lauded in chronic bronchial affections. One of the simplest is suggested by Pasch. It consists in putting a drachm of liq. ammoniæ in a saucer, and placing in this a watch-glass with about half a drachm of pure muriatic acid. White fumes of muriate of ammonia arise, which may be inhaled through a paper funnel.

Of all the gases oxygen is now being most tried, particularly for the relief of dyspnœa and in low fevers, or during convalescence from low fevers, and in chlorosis. We must await these new trials; for the former experiments were unsatisfactory. But excellent results have quite lately been claimed for oxygen inhalations by Demarquay†—who is at present investigating the subject—not only in the conditions named, but in diabetes, in senile gangrene, and in prolonged suppuration. The gas is inhaled pure; and may be generated by any of the processes known to chemists. Several apparatuses for its convenient and speedy production have, however, been recently brought forward; and one of the kind has been patented by Dr. Beigel. The apparatus made by Galante in Paris is very well arranged.

---

* Transactions of the Am. Med. Association, vol. iii.

† See Reports to Acad. of Medicine; and Gazette Medicale, 1866.

CHAPTER V.

THERAPEUTIC CONSIDERATIONS.

As the mode of using inhalations and the general questions connected with their employ have now been considered, we may turn to the more strictly clinical part of the inquiry, and examine, by the light of observation, into their real value as therapeutic means. In so doing I shall only be able to discuss the applicability of the atomized fluids to the treatment of some of the principal diseases of the respiratory organs; for to do otherwise would be to write a treatise on respiratory therapeutics rather than an essay bearing on the subject. Nor can I introduce in detail all the material I have collected even on the matters brought forward, but shall allude only to such parts of it as are the most significant, embodying, however, all in any deductions made.

And first, to look at *laryngeal affections.* Here I have found atomized liquids of most service in the catarrhal conditions, whether connected with pharyngitis or not. In the loss of voice and irritative cough associated with *catarrhal laryngitis*, not, however, while in its most acute stage, I have repeatedly known inhalations of alum, about ten grains to the ounce, combined with five to six drops of laudanum, or with conium, produce a most happy and soothing effect, and exert sometimes an almost immediate influence on the voice. Thus, in a case of ten days' standing, in an elderly gentleman, the voice became after each inhalation, two to three being employed daily, quite distinct, though at first it resumed its whispering tone between them. In a few days a permanent result was perceptible. I have obtained an equally good result from a solution of chlorate of potassa. The soothing effect of the remedy in one case particularly in which a high degree of both pharyngeal and laryngeal congestion of upwards of a week's duration existed was very marked. Inhalations of pulverized warm water, with or without narcotics, are very grateful in the acute or subacute catarrhal conditions, since they relieve much the sense of dryness and of heat.

In the more chronic cases, and when marked swelling of the mucous membrane of the larynx and vocal cords is perceptible, the effects of inhalations of atomized fluids are not always quickly evident; but in these cases, too, I have used alum, tannin, sulphate of zinc, and the subsulphate of iron, with advantage, commencing with small doses. For instance, I was consulted by a clergyman who had strained his voice by incessant speaking, and had in addition caught a severe cold. His voice, from having been one of remarkably

fine compass, had become coarse, and was at times so
hoarse as to be whispering. The laryngeal mucous
membrane was much congested, and there were a few
rales in the chest from accompanying bronchitis. He
coughed much, complained of an uneasiness in the
throat, and was rather short of breath; otherwise no
symptoms of any disorder existed. He had tried for
three or four weeks various internal medication, without benefit. I directed him inhalations of tannin, five
grains to the ounce, subsequently somewhat increasing
the strength. After the sixth inhalation his voice improved most strikingly, and it continued to improve for
ten days, at which time he left the city. When not
endeavoring to speak too loud, the voice was quite
clear, and had nearly regained its natural tone; the
cough had almost ceased.

From the subsulphate of iron I have seen similarly
good and even prompter effects. Not long since I
employed it in a case in which great swelling of the
epiglottis existed, concealing to a very considerable
extent the structures within the laryngeal aperture,
and attended with much difficulty in swallowing and
aphonia. The disorder had lasted for more than two
months, and the loss of voice had been gradually progressing, until, five days before I saw the patient, the
voice had been reduced to a mere whisper. Several
drachms of a solution of Monsel's salt, sixteen minims
to the ounce, were injected by a hand-ball atomizer,

and before the young man left the office his voice was distinctly audible. He came back two days afterward, speaking quite plainly, and stating that he had been able to swallow solid food, the first for weeks. The inhalation was repeated, and both voice and power of deglutition again markedly improved. Subsequent examination with the laryngoscope showed the most evident reduction in the tumefaction and change in the color of the engorged structures.

In the various forms of *ulceration* of the laryngeal structures, the method of treatment under discussion has been applied by means of pulverized solutions of tannin, of corrosive sublimate, of iodine, of iodide of silver and of nitrate of silver. I have used tannin, sulphate of copper and nitrate of silver in several cases, but have not obtained good results. For example, in a gentleman, forty-two years of age, on whose right false vocal cord the laryngoscope detected an ulcer with irregular borders, inhalations of the compound solution of iodine, commenced with 10 drops to the ounce and gradually increased, were faithfully tried for fully three weeks without any perceptible benefit being produced. Subsequent touching with nitrate of silver, the hand being guided by the reflected image of the parts in the laryngeal mirror, proved far more effectual, both in the improvement manifest in the ulcer and in the symptoms of impairment of voice, difficulty in swallowing, and cough. The ulcer, judging from the history of the case, was probably scrofulous. In the following case of laryngeal ulceration the treatment by inhalation was also fully tried:

A man, forty-five years of age, was admitted into the Pennsylvania Hospital on the 13th of February, 1866, with a cough which he traced to exposure dating eight weeks back, though when questioned he stated that he had a very slight dry cough, off and on, for a month previous to this. The severe cough accompanying the cold he had caught was attended with sore throat, and soon afterward with hoarseness. On admission he was noted to be pale and to present a sickly aspect; respirations 24; expectoration but slight in quantity, tough and whitish; and neither fever nor deficient appetite. No abnormal physical signs were discernible in the lungs, save a slight harshness at the lower part of the left. There was difficulty in swallowing, without pain on pressure over the larynx; the voice was hoarse, but not completely lost; the fauces were not reddened. On laryngoscopic examination a large superficial, yellowish ulcer was seen on the right false vocal cord, extending to the arytæno-epiglottidean fold. There was also considerable thickening of these structures as well as those of the left side, but the true cords seemed unaltered and approximated fully in the act of vocalization. The man denied the existence of any syphilitic taint.

On the 26th of the month, having since his admission

had his larynx touched several times with nitrate of silver, besides taking tonics, he was directed to use daily inhalations of alum, 30 grains to the ounce. On the 9th of March this treatment was stopped, and he complained of his throat feeling very sore—an occurrence which, with a sense of oppression and tightness and an aggravation of the cough, I have several times noticed from the use of very strong solutions of alum. Finding, on examination with the laryngoscope, ulceration beginning on the other cord, and perceiving no amelioration in any of the symptoms, the alum solution was not resumed, but sulphate of copper inhalations, $2\frac{1}{2}$ grains to the ounce, were substituted. On the 18th, as they had produced no effect on the ulcers, though the swelling was less, the strength was doubled, and he inhaled an ounce of the solution daily without any inconvenience, though twice it made him sick at the stomach. His general condition was not satisfactory, and in addition to iodide of iron and an anodyne cough mixture, he was placed on cod-liver oil, a tablespoonful three times daily. A few days afterward an examination of the chest showed coarse, dry rales in expiration in both lungs, and a more high pitched percussion note with greater resistance at the right apex. The coarse, dry rales were there, too, more distinct; there was more cough, followed by a frothy and copious expectoration; and, altogether, it was evident that a tubercular infiltration into the lung was taking place.

6*

From this time on the history was that of a well-developed case of phthisis. He had much cough and profuse expectoration, with rapidly progressing emaciation, and then night-sweats. The dry rales gradually disappeared, giving way to harsh breathing, and a month after the date last mentioned signs of softening were clearly discernible at the right apex.

But to return to the laryngeal symptoms and, their treatment by inhalations. The inhalation of copper, which was kept up until the 26th of March, considerably lessened the frothy expectoration and somewhat reduced the swelling, but it did not put a stop to the progress of the ulceration. Nitrate of silver, with a brush, was then several times used, and on the 7th of April two fluid drachms and a half of a solution of nitrate of silver were administered by means of the hand-ball atomizer. The injection produced a burning sensation, lasting two hours, and a marked abatement in the cough. But on the 9th the ulceration, instead of decreasing, was found to have extended to the right true vocal cord, which was decidedly excavated on its margin; a few isolated, yellowish spots were also seen on the wall of the trachea. The poor man had much difficulty in swallowing, but had not completely lost his voice.

From the 9th to the 16th of April, he took, with the steam atomizer, six inhalations of nitrate of silver, half an ounce at a time, of the strength of fifteen grains to the ounce. Twice, after inhaling, he was sick at his stomach. Subsequent to each inhalation, it was noted that his larynx smarted for an hour; but for a few hours the expectoration ceased. From this period on until he left the hospital, April 30th, and very shortly before his death, the inhalations were not regularly kept up. A few local applications were made by means of a sponge, and he expressed himself always as being relieved by them. His increasing weakness caused him to prefer them to inhalations. But nothing really gave him much relief; the difficulty in swallowing was so great that he had to be nourished exclusively by fluids; there was tenderness on pressing between the hyoid bone and larynx; the cough was very annoying, and the sputa, no longer so frothy, were obviously nummular; the voice was reduced to an almost inaudible whisper. The last laryngoscopic examination, made after the inhalations had been stopped, showed that the ulceration had greatly altered the true cords. The false were less swollen, and the ulcer on them had not increased, but an ulcer was also seen on the outer edge of the left arytæno-epiglottidean fold.

In reviewing this singular case, we are struck with the sudden beginning of the affection in the larynx and with the laryngeal phthisis, preceding that of the lungs. But this point of the case cannot be here discussed. I have introduced it rather to study the effects of the inhalations; and though these were, on the whole, of some service in reducing the swelling, and though thus we may claim that a certain degree of comfort was procured, it cannot be said that either the sulphate of copper or the nitrate of silver arrested the extension of the ulceration. Nor were the results obtained by the latter agent greater than, indeed not so great as, those produced by the local application of nitrate of silver with a sponge or brush. In simple ulcers, inhalations may be of more decided use; though even here I much prefer, so far as I have tested the matter, the other method of local treatment.

In *œdema of the glottis* tannin has been greatly lauded by Trousseau. In the first volume of his Clinique Médicale he records a case in which a strong solution of tannin was inhaled every hour, with the most obvious effect on the attacks of suffocation, and indeed on the disease. During the second day there was but one fit of suffocation, and the respiration had lost its noisy character. The attacks recurred once in twenty-four hours for three days, but on the fourth day of treatment the respiration was natural. The young woman left the hospital a few days afterward, perfectly convalescent.

The same treatment, too, proved of service in the hands of Barthez,* at the Children's Hospital, St.

---

* Traitement des Angines Diphthéritiques par la Pulverization. Paris, 1861.

Eugéne, in *laryngeal diphtheria* and in *croup*. He
cites four cases in which a tannin solution, from five to
ten per cent. strong, was inhaled from eight to twenty
times in the course of twenty-four hours, each inhalation lasting from fifteen to twenty minutes, and being
always followed by evident temporary relief. Two of
the children recovered, the other two died. But the
autopsy proved that the false membrane had entirely
disappeared. Death was due to the diphtheritic poison-
ing. The results of the tannin inhalation are attributed
by Barthez to the astringent effects of the tannin on the
membrane, which, when corrugated, rolls up at the
edges, and is thus prone to be gradually detached.
But Fieber, who treated fifteen cases with tannin
solution much in the same manner, and who reports
ten cures among them, attributes the success to the
dissolving influence of the remedy. Yet, when we
come to examine critically the instances recorded by
Barthez, the former supposition becomes far the more
probable. Thus, in analyzing his cases, I find that the
first was sick for five days before admission, and seven
days under treatment, making, so far as can be judged
from the record, from nine to ten days that the membrane lasted. In the second case the treatment did not
begin until the second day of the sickness; the child
died on the twelfth day. In the third case, which seems
to have been one of pseudo-membranous croup rather
than of laryngeal diphtheria, the treatment commenced

on the fourth day of the malady, and by the ninth day
the little patient had recovered. The fourth case was
eight days sick when the treatment by inhalation began,
and was subjected to it for four or five days before full
recovery took place. Now this does not look like any
marked solvent power of the remedy, for diphtheritic
membranes are not permanent structures, but are very
apt to disappear from the circumference to the centre
within a week after their appearance. Hence, if we
accord any value to the treatment—which, bearing in
mind the usually fatal character of laryngeal diphtheria
and the grave character of pseudo-membranous croup,
we cannot totally refuse to do—we must also admit
that the action is not rapid, and not what we might
expect from a solvent. Nor can we overlook the effect
of the water in the combination as a cleansing agent,
and as tending to aid in removing and in expectorating
the breaking down textures; for Siegle* used inhalations of pulverized warm water alone in a case, apparently hopeless, of membranous croup, with the greatest
relief to the child, and with the result of causing it to
expectorate with the dense mucus shreds of the membrane. The child recovered.

But to return to *diphtheria*. Other agents besides
tannin have been resorted to, to counteract its local
manifestation, both when the larynx is implicated and

---

* Hals und Lungenleiden, 1865.

when the membrane has not extended to it—chloride of iron, chlorate of potassa, alum. Lewin* has reported at length fifteen cases, eleven of which recovered. I have analyzed these cases, and find the following result. Only in four of the eleven that recovered (Cases IV., VI., VII., and XI.) was the larynx implicated, and in these four the membrane was chiefly on the upper surface of the epiglottis. Only in one, Case XI., did it extend to the under surface of the epiglottis and to the arytenoid cartilage. Of the four fatal cases the larynx was in three very decidedly affected. This result is therefore by no means remarkable, particularly as cauterizations and internal remedies were at the same time used. Indeed Lewin himself speaks more of the action of the inhalations in preventing the membranes from re-forming than of their power to remove those already formed.

Yet another remedy that has been recommended, both in diphtheria and in croup, is lime-water. Küchenmeister and others have found that the pseudo-membrane was soluble in concentrated acetic acid, in alkalies, in carbonate of lithia, but with greatest readiness in lime-water, and the attempt has been made to employ this by atomization as a solvent. Biermer† was the first to use it. The patient was a girl, aged seven-

---

* Inhalations Therapie, 1865.
† Schweizersche Zeitung für Heilkunde, 1864.

teen, admitted into the hospital at Berne for croup, which had lasted for four days; the suffocative phenomena were very marked. To moisten the respiratory passages pulverized water was tried, first warm, then boiling. After inhaling for an hour, with much comfort, vehement coughing occurred, whereby a quantity of mucus and shreds of false membrane were discharged, causing decided relief. This became still more evident when warm lime-water (one part of lime to thirty of water) was used with the nebulizer every second hour, each inhalation lasting a quarter of an hour. Thick, purulent matter and crumbling pieces of membrane were expectorated, and the signs of laryngeal obstruction gradually disappeared.

Biermer insists on the necessity of using the inhalations hot. Dr. Küchenmeister and Dr. Brauser* have each published a successful case treated in the same manner, and a substitute for the pulverized inhalation has been attempted in this country by Dr. Geiger, who poured hot water on unslacked lime, and caused the steam arising from it to be inhaled. He reports several cases of pseudo-membranous croup with a fortunate issue.†

Not having tried atomized lime-water in croup or in laryngeal diphtheria, I cannot speak from personal

---

* Referred to in British and Foreign Med.-Chir. Review, July, 1865.

† Medical and Surgical Reporter, April, 1866.

experience either of its effect or want of effect. But I have watched in two cases of diphtheria, with some care, the action of lime-water on the visible deposits.

In the first case, that of a lady, seen in consultation with her physician on the fifth day of her confinement, the deposit covered the roof of the mouth, the half arches and part of the wall of the pharynx. There was also, and indeed the progress of the case placed the matter beyond doubt, reason to believe that nasal diphtheria existed. She was taking chloride of iron, full nourishment and stimulants. I directed a stream of pulverized lime-water about eight times stronger than that officinal in our pharmacopœia—the liquor calcis saccharatus of the British pharmacopœia—on the affected part, by means of an excellent hand-ball atomizer, for three or four minutes at a time. This treatment was carried on every few hours, but no perceptible influence on the membrane could be detected. The application was cleansing and very grateful, particularly so when thrown up the nostril. And here let me, in passing, state that the spray was felt to arrive in the throat, and that though the remedies we resort to may not succeed in dissolving the membrane, I beg to draw attention most earnestly to this use of the atomizer in nasal diphtheria, and particularly in the nasal diphtheria of children, as an excellent means of acting locally on the affected part. But to return to

7

the case. It terminated fatally, the membranes in the mouth remaining in a very thick layer.

The second case was that of a gentleman, thirty-five years of age. Here there was no nasal diphtheria, nor were the constitutional symptoms by any means so grave; and after the disappearance of the membranes, which took place in about nine days, convalescence was rapid. As local treatment, early in the affection, a strong solution of sulphate of copper was employed. But I also, both at the time and afterward, made use of atomized solutions of lime, in the same manner as in the preceding case, and not hot. The remedy was again very grateful and cleansing; yet, though I selected repeatedly the same spot on the left half arch to throw the solution on, I could not see that it had any perceptible effect in thinning the deposit.

If lime-water, then, be a solvent of the membrane on living textures, it is so very gradually, and much of it would have to be employed to produce a decided result. In cases running an acute course it could therefore not be depended on. Indeed, to take Prof. Biermer's case as an example, how much may have been due to the inhalation of the warm fluid alone? That warm water by itself is serviceable we know from Siegle's case, already mentioned.

In *hooping-cough*, Dr. Steffen* claims to have had

---

* Journal für Kinderkrankheiten, Jan. and Feb., 1866.

success with an inhalation of common salt and opium, and with five grains of tannin and three drops of laudanum in two ounces of water, used daily. But though the inhalations seemed to have afforded some comfort to the children, the duration of three of the cases was from two to three weeks, and one case lasted nine weeks. Hyoscyamus, alum, and perchloride of iron have also been employed in this disorder.

In *asthma*, arsenic has been recommended by several physicians, Trousseau and Eck among them. Chloride of sodium with laudanum has also been employed; and I have used lobelia and conium in several instances. All these remedies except the arsenic are chiefly resorted to at the time of the paroxysm. But I cannot say that I have seen anything produce decided results. The inhalation of pulverized warm water alone has seemed to me quite as grateful to the patient as medicated solutions, and has promoted expectoration.

Turning now to *pulmonary affections*, we shall inquire into the effects of inhalations on bronchitis, phthisis, and hemorrhage from the lungs. In acute *bronchitis*, inhalations of pulverized warm water will often afford much comfort and assist in inducing expectoration. Medicated inhalations I have not used, nor do they seem to have been much resorted to. But in chronic bronchitis it is evident that they have a large field for employ, and the results are sometimes very striking. To cite a case.

Mrs. ——, 48 years of age, consulted me in January for a cough which she had had for upwards of a year and a half. Getting much better during the earlier summer months, she had, in the autumn of last year, after severe and prolonged exposure to wet, a violent bronchitis, or broncho-pneumonia, attended with probably much pulmonary congestion, and shortly followed by a hemorrhage. She stated that her cough was very distressing; the expectoration purulent, profuse, having an unpleasant odor. Moreover, it had contained blood daily, for several months, varying in quantity from mere streaks to an amount which gave to the whole sputum in the cup a decidedly bloody appearance. Her breathing was oppressed; at times so much so as to be wheezing and asthmatic. Examining the lungs, they were found to be filled with rales, dry and moist, the latter far preponderating. There was no decided dulness on percussion, though at the upper part of the right lung an impairment of resonance existed, which may have been due to the partial consolidation of the pulmonary tissue occurring at the time she had the attack of bronchitis or broncho-pneumonia, subsequent to the exposure alluded to. The appetite was good; the general health excellent. As she had tried many remedies faithfully, some under the advice of an eminent physician, I determined to use atomized fluids by inhalation, and directed an ounce of a solution of alum, 15 grains to the ounce, with 6 drops of fluid extract of conium.

The first inhalation produced not only no relief, but a decided constriction in the chest; the second, taken the next day, led to a severe paroxysm of difficult breathing. Finding the astringent action of the remedy too great, I reduced its strength to about 8 grains to the ounce. She bore this perfectly well, and after the third inhalation, counting in the first two, a change in the sputum was noticeable. It was somewhat less copious, and the quantity of blood in it was obviously diminished. She took altogether, while in Philadelphia—of which she was not a resident— nine inhalations, and when she left the city, though the expectoration was still very copious and as yet but slightly changed in character, it no longer contained a trace of blood.

During her stay here little or no internal treatment was employed; but on leaving, while urging her to continue her inhalations by the atomizer, and to vary them at times by breathing the vapor of tar, I also directed her chest to be painted with croton oil, and gave her a cough mixture, of which wine of tar and fluid extract of wild cherry formed the chief ingredients.

This treatment was carried on for fully two weeks, when I was informed, by letter, that she had used the common inhaler with tar and warm water twice each day, that she had finished her cough mixture, had employed the alum inhalation daily, and that she still

7*

had "slight turns of wheezing on lying down, which lasted from half an hour to an hour, but none so bad as that one in Philadelphia, and they are somewhat relieved by inhaling hot camphorated water. The expectorations are all, or nearly all, from 4 or 5 A.M. to 10 A.M. After that there is but little cough. The expectoration last week was for a day or two offensive, but is very little so now. It is lighter colored, and has not been at all bloody since we returned."

After this, for six weeks daily, she went on steadily with the alum inhalations, increasing the strength to 20 grains or somewhat upwards to the ounce of water. She also resorted occasionally to the ordinary tar inhalations alluded to, and at times to pulverized solutions of common salt. The cough medicine was abandoned and wine of tar taken, though this was not persevered in; the alum inhalations were her main dependence. Under this treatment she gradually recovered: the cough and all expectoration ceased; the asthmatic seizures no longer took place, and when I saw her in Philadelphia, in April, she had been for several weeks perfectly well. She had at that time some rales in her chest, and a very slight expectoration from a catarrhal condition of a few days' standing, but otherwise she presented all the signs of good health. The partial dulness under the right clavicle had all but disappeared.

This case is certainly very interesting as regards the

use of inhalations. The unfavorable consequences at first from the too strong solution of alum employed; the speedy disappearance of blood from the sputum; the gradual cessation of the expectoration; the slight general treatment made use of—and to no portion of which does it appear that any decided importance can be attached—are all decided features in the case. And though we may affix some value to the inhalations of chloride of sodium and of tar, yet it is evident that the most efficient remedy was the alum.

Besides this remedy, good results may be obtained from the use of tannin, of sulphate of zinc, of iodine, and, where we wish to promote the expectoration, of muriate of ammonia or of chloride of sodium—to all of which a small quantity of a narcotic solution can be serviceably added. Yet these agents are not always of advantage. The extent of the alteration of the mucous membrane has a great deal to do with the success of the treatment. I have a gentleman under my care who has had chronic bronchitis, with excessive secretion, for twelve years, scarcely influenced by the various climates which he has sought. In his case inhalations of iodine, of tannin, of carbolic acid, of lobelia, of sulphate of zinc, of alum, of muriate of ammonia, of chloride of sodium, have thus far been used to little if any purpose.

There is no disease for which inhalations are more likely to be eagerly resorted to than *phthisis*. I have employed them and noted their effects with care in quite a large number of cases; but it is impossible here to give more than the general results, and in so doing it will be convenient to separate the effects of the inhalation on the disease itself and on its more prominent symptoms. I will take for analysis ten cases, treated at the Pennsylvania Hospital, and in which either no other remedies were prescribed, or merely remedies to fulfil a temporary indication. In four cases chlorinated soda was used, the liquor sodæ chlorinatæ of our pharmacopœia. It was employed every day, in doses varying from half a drachm to a drachm to the ounce; or sometimes two inhalations were administered daily, of half a drachm each. In the doses mentioned it was perfectly well borne, and although at first it irritated and had to be given in a more diluted form, after a few days it was taken without difficulty. In the first case— a case in which softening was just beginning—the inhalations were used daily for eighteen days. They caused no difference either in the physical signs or symptoms, though the patient stated that he coughed less and that the expectoration was much easier after them. In the second case the effect on the cough was similar, and the sense of tickling in the throat, for which the patient had previously tried several remedies in vain, was quickly relieved. A decided improvement also took place in her general condition; but the same dulness under the right clavicle, with crackling, which existed at the beginning of the treatment, was found after she had for twenty-two days inhaled daily a solution of the strength of one drachm to the ounce. On the whole, however, the remedy appeared to have a beneficial influence. In the third case twenty-five daily inhalations were used. The cough improved. The disease, which had not advanced to softening, remained stationary. But neither physical signs nor general condition showed any decided amelioration. In the fourth case the chlorinated soda was not used very long and produced no appreciable effect.

The results in these cases were not particularly encouraging, though not totally negative. Two cases were then treated with iodine inhalations; liquor iodinii compositus, viii minims to $\mathfrak{z}j$ increased to xv minims to $\mathfrak{z}j$ taken daily or twice daily. Both improved—one strikingly. This was a case of tubercular disease of both lungs, following right-sided pleurisy. There was crackling (not, however coarse) at both apices, and he was losing flesh and strength rapidly, notwithstanding that he was taking cod-liver oil and iron. The cough was dry and irritative. He used every day, for a month, iodine inhalations, at first eight minims in each, then fifteen minims morning and evening. The internal treatment was stopped. He gained several pounds of flesh; his appetite became good; the respirations came down to 18, the night-sweats ceased, and an undoubted change took place in the physical signs; the crackling almost disappeared, the dulness lessened. The inhalations at first produced some irritation and a little cough. They were always followed by slight expectoration.

Solution of chloride of iron was used in two cases. In the one there was coarse crackling, with distinct blowing and rather low-pitched respiration under the right clavicle, and coarse crackling on the left side; the symptoms were those of phthisis passing into the stage of softening. One-eighth of a grain of perchloride of iron was used morning and evening for sixteen days. It rather reduced the expectoration, but did not influence the progress of the disease. In the second case the effects were decidedly beneficial. When admitted into the hospital there was dulness on percussion, with harsh breathing under the left clavicle; hacking cough of several months' standing; great pallor and marked anæmia, which may, however, in great part at least, have been due to a severe attack of menorrhagia. The pulse was 108; the respirations 26. The perchloride of iron was administered in the same way as in the preceding case. In a week after she had commenced it, the lips were of far better color, and she began to gain flesh and strength. The dose was, after seventeen days, increased to one-half a grain daily, which she took for ten days, all the time improving. For a short period a solution of pyrophosphate of iron was substituted, but she went back to the chloride.

The iron inhalations were used for about six weeks, and shortly before leaving the hospital her cough had all but disappeared. There was a scarcely appreciable difference in the percussion note between the two sides of the chest, and the respiratory murmur had lost its harshness. She did not feel quite so strong and well as ten days previous, in consequence of an attack of in· termittent fever. I have heard that since she left the hospital she has had two hemorrhages.

In the two remaining cases out of the ten, chloride of sodium and muriate of ammonia were chiefly employed. They were not without influence on the cough, and on the expectoration; but were apparently no check to the disease.

Thus it will be seen that iodine and iron inhalations both had a decided effect where softening had not as yet occurred. But did they do so in virtue of any local action, or of their general power on the economy after being absorbed by the respiratory mucous membrane? This question is one difficult to solve, save by the most careful observation of a large number of cases. But under any circumstance it would certainly seem that these remedies merit a trial in cases of early pulmonary tuberculosis. Supposing the inhalations to be well borne and rather comforting to the patient, as they mostly are, do we not then introduce desirable medicines into the system without inconvenience, and by carefully making use of the lungs, save the stomach?

The foregoing statements represent the analysis of the ten cases, treated under circumstances permitting of their careful study; but they also represent my impressions derived from a far larger number of cases, of which, however, other treatment being at the same time employed, so rigorous an analysis could not be made. I may also add that I have used carbolic acid by atomization in the treatment of phthisis; yet have not seen that it in any way arrested the disease. It was rather grateful to the patients, gave them an increased feeling of comfort in breathing, and had an influence in promoting expectoration, and was therefore useful; as indeed it may be even when employed by the ordinary inhaler in early phthisis or in chronic bronchitis. But I have never seen a case of consumption arrested by it.

In turning to the symptoms of phthisis we find, as regards the cough, especially when occurring in paroxysms, that pulverized solutions of the watery extract of opium, of conium, or of cannabis indica afford relief; and these remedies are particularly serviceable when the cough produces vomiting, or is associated with great gastric irritability. In cases of cavities with purulent contents, or under any circumstances to render the sputum more easy of expectoration, solutions of common salt or of muriate of ammonia are of avail. Where the sputa are copious and offensive, tar has been recommended. But judging by the case of a gentleman whose lungs were riddled with cavities, and whose expectoration was purulent, profuse, and very fetid, tar inhalations pursued in the ordinary manner are better, certainly quite as well, borne, and afford more comfort.

As regards *hemorrhage from the lungs*, the evidence that has been collected in favor of the treatment by atomized liquids appears very decided. A number of cases have been reported by Sales Girons, by Lewin, by Siegle, by Zdekauer and others, in which, instantly after inhalations of strong solutions of alum, or of chloride of iron, the hemorrhage stopped. I have used alum, chloride of iron, and persulphate of iron, and have thought that the remedies had so distinct an effect that I should not abstain from resorting to them in any case of pulmonary hemorrhage at all unyielding. At the same time, as the cases in which the inhalations were employed were on active internal treatment, I do not bring them forward. There is nothing more difficult to establish than the relation between cause and effect in hæmoptysis. The *post hoc propter hoc* is here so uncertain that evidence, to be accepted, ought to be of the most unimpeachable kind. Of my six cases this cannot be said. I will, however, state that, contrary to what may be expected, the inhalations gave rise to no irritation nor coughing or oppression. They may be used very strong. In one of the cases alluded to, one of our Resident Physicians at the Pennsylvania Hospital, Dr. Herbert, employed a saturated solution of alum; in another, a drachm of Monsel's solution of subsulphate of iron to an ounce of water. The cases under discussion were all extremely severe. In mere spitting of blood, or in instances of blood-streaked sputum, the influence of the astringent remedies are often very obvious. The case of the lady above reported, affected with bronchitis, is a case in point. I have seen one quite as striking, in which two inhalations of a solution of subsulphate of iron completely and permanently arrested a bloody expectoration which had lasted for four months.

Summing up, now, the results of this inquiry, it may be stated to lead to these conclusions :

1. That inhalations by means of atomized fluids are an unquestionable addition to our therapeutic means; but that they are nothing but an addition, and not a substitute for all other treatment; that therefore their claims to be so considered are unfounded.

2. That in most acute diseases of the larynx, and still more so in acute disorders of the lungs, their value, save in so far as those of water may tend to relieve the sense of distress, etc., and aid expectoration, is very doubtful; though in some acute affections, such as in œdema of the glottis and in croup, medicated inhalations have claims to consideration.

3. That in certain chronic morbid states of the larynx,

particularly those of a catarrhal kind, and in chronic bronchitis, they have proved themselves of value; but that they are useless or next to useless in ulcerated diseases of the larynx.

4. That in the earlier stages of phthisis they may be of decided advantage, and that at any stage they may efficiently aid in treating the symptoms of this malady; but that they are valueless to stay the disease after softening has fairly set in.

5. That their influence on such affections as hooping-cough and asthma is not satisfactorily proven.

6. That they furnish an unexpected augmentation of our resources in the treatment of pulmonary hemorrhage.

7. That the question in any disease of the respiratory tract is not whether the atomized fluids can reach the seat of the malady, but whether they can do so in sufficient quantity, and in a manner to become available as a therapeutic means.

8. That in estimating the action of inhalations of atomized fluids, we must accord due value to the ready absorption of many through the pulmonary structures, and guard against attributing to a local influence what may be due to the constitutional effect of the remedy.

9. That we cannot overlook the part the watery vapor plays when using atomized solutions.

10. That they require much care in their employ; and that particularly in acute affections we should consider

whether, as they have to be used frequently to be of
service, the patient's strength justifies the disturbance
or the annoyance their frequent use may be.

11. That in any case, to be of service, the inhalations
ought to be carried on as a treatment with a distinct
object, and not intermittingly or spasmodically resorted to.

These conclusions and the remarks preceding them
apply exclusively to the treatment of the diseases of
the respiratory passages by atomized fluids, for though
incidentally the inhalation of gases or vapors has been
mentioned, it has purposely been no more than alluded
to, since this subject has been long before the profession and has been often examined; whereas that of
inhalations by means of atomized fluids is a novel one,
and one which will require much unbiased investiga-
tion to determine its true position. Nor has the applicability of atomized fluids to affections of the fauces
been here particularly brought forward; though, as
previously indicated, they are of striking value in these
affections. In many a case of chronic sore-throat I
have used astringent solutions, as tannin, alum, or solutions of chlorate of potassa, or sulphate of copper, with
decided success. In acute sore-throats, or in affections
of the mouth, sedative and anodyne solutions, or pulverized ice water has often proved most grateful.
And I have seen elongated and tumid uvulas which
would formerly have tempted any surgeon to snip

them off, yield to the use of astringent solutions thrown directly on them in the form of spray.

When, now, we contrast inhalation by atomization with other kinds of inhalation, we find that by its means we can use substances, such as astringents and caustics, which were formerly not available. Thus this method has greatly extended the range of inhalations. Moreover, there is very much greater certainty of a local action than is otherwise obtainable. As regards laryngeal diseases, however, the local application of remedies, guided by the laryngoscope, is on the whole more certain and efficient.

In conclusion, I will point out what a wide range of applicability atomization has beyond that to the treatment of the diseases of respiration, or even of the fauces, nares, or any part on which a local action is desirable. By atomizing salt or iodine in rooms or in wards of hospitals, we can cause our patients to breathe constantly an atmosphere impregnated with these agents, if such an atmosphere be thought desirable. Permanganate of potassa, chlorine, or carbolic acid may be used in the same manner as disinfectants; and, as I have tested now in many instances, we may obtain by atomization the constitutional effects of remedies on the system. In cases in which the stomach cannot be resorted to this will be a great aid. It is scarcely necessary to dwell on its value, for instance, in anæmia, with enfeebled digestive powers. The

effect of remedies, too, thus administered, is generally
very prompt. I have seen the pupils dilate and a
staggering gait produced by breathing for a few minutes a pulverized solution of conium. Of course it is
incumbent upon us not to avail ourselves of the respiratory mucous membrane needlessly; and if it be
employed, it ought to be done so with care, for it is not
a membrane that will bear the slights and rude usage
the stomach receives. But it is a great satisfaction to
know that, should we wish to make use of the lungs to
introduce medicines into the system, we possess now a
means more certain, more efficient, and more susceptible of being regulated than any that was formerly
available.

www.ingramcontent.com/pod-product-compliance
Lightning Source LLC
Chambersburg PA
CBHW021955190326
41519CB00009B/1262